COMPUTER NETWORKING FIRST-STEP

A Beginner's Guide to Understanding Computer Architecture and Mastering Communications System Including Cisco, CCNA, CCENT, and the OSI Model

© Copyright 2019 - All rights reserved.

This content is provided with the sole purpose of providing relevant information on a specific topic for which every reasonable effort has been made to ensure that it is both accurate and reasonable. Nevertheless, by purchasing this content you consent to the fact that the author, as well as the publisher, are in no way experts on the topics contained herein, regardless of any claims as such that may be made within. As such, any suggestions or recommendations that are made within are done so purely for entertainment value. It is recommended that you always consult a professional prior to undertaking any of the advice or techniques discussed within.

This is a legally binding declaration that is considered both valid and fair by both the Committee of Publishers Association and the American Bar Association and should be considered as legally binding within the United States.

The reproduction, transmission, and duplication of any of the content found herein, including any specific or extended information will be done as an illegal act regardless of the end form the information ultimately takes. This includes copied versions of the work both physical, digital and audio unless express consent of the Publisher is provided beforehand. Any additional rights reserved.

Furthermore, the information that can be found within the pages described forthwith shall be considered both accurate and truthful when it comes to the recounting of facts. As such, any use, correct or incorrect, of the provided information will render the Publisher free of responsibility as to the actions taken outside of their direct purview. Regardless, there are zero scenarios where the original author or the Publisher can be deemed liable in any fashion for any damages or hardships that may result from any of the information discussed herein.

Additionally, the information in the following pages is intended only for informational purposes and should thus be thought of as universal. As befitting its nature, it is presented without assurance regarding its prolonged validity or interim quality. Trademarks that are mentioned are done without written consent and can in no way be considered an endorsement from the trademark holder.

Table of Contents

Introduction
Chapter 1: What is Computer Networking?................5

    Computer Network Classifications..6
    Network Design..7
    The Network Protocols..8
    The Hardware and Software Needed...9
    Home Computer Networking..10
    Business Computer Networks...13
    Other Things to Consider About our Network.........................14

Chapter 2: The Main Components of a Computer.......18

    The Input Unit...20
    Memory or the Storage Unit...22
    Arithmetic or Logic Unit...23
    Output Unit...24
    The Control Units...24

Chapter 3: The Different Types of Computer Networks..28

    The Local Area Network...29
    The Personal Area Network..30
    Metropolitan Area Network..32
    Wide Area Network..34

InternetWork..39

Chapter 4: What is the Virtual Memory?....................43

Chapter 5: A Closer Look at the Networking Infrastructure..48

Chapter 6: Understanding the Protocols of a Network...55

Internet Protocols...57
Wireless Network Protocols..57
Network Routing Protocols...59
Other Protocols to Consider:......................................59
How These Network Protocols are Implemented...................64

Chapter 7: How to Handle IP Addressing....................67

What is the IP Address?...69
The Subnet Mask..72
The Default Gateway Address....................................73
The DNS Servers..75

Chapter 8: A Look at the CISCO Networking Technologies...78

Why Machine Reasoning is Better than Machine Learning..80
Optical Networking..84
CBRS and Cognitive Radio Emerging Networking Technologies..87

Chapter 9: How to Handle CCNA and CCENT.............90

Chapter 10: Introducing the OSI Model........................98

What is the OSI Model?..99
The Seven Layers of the OSI Model?.....................................101

Chapter 11: A Word About Network Security..............106

Why is Network Security Important?....................................107
What Can Go Wrong On My Network?................................109

Conclusion..114

Introduction

Congratulations on purchasing *Computer Networking – First Step,* and thank you for doing so.

The following chapters will discuss everything you need to know to get started with some of the computer networking that we need to use for our needs. There are many parts that are going to come into play when it is time to handle our own computer network. And understanding what all goes into a computer network, why it is all so important overall, and more, is going to make a big difference in the amount of success we can have with making our network behave in the manner that we would like.

This guidebook will spend some time talking about the importance of a computer network, and what it is all about. We are going to spend our time looking at how we are able to set up a network, what is going to be found in one of these networks and more. We will then take it a bit further and look at some of the basic parts of a computer that are important to our network, before diving in to some of the most common computer networks that are available for us to use based on the size that we would like the network to be.

We will also expand out to a few more topics throughout this guidebook. For example, we will look at how we can work with the IP addressing and how this is important when it is time to put together the communications with others, some of the security protocols that are out there, and how to choose one based on the needs that you have, and how to pick from some of the best Cisco technologies that will bring your work to the next level.

From there, we will explore a few more things that are important to helping us to understand our network and ensuring that we are able to get it set up in the manner that we would like. For example, we can finish out this guidebook with a look at what CCNA and CCENT mean and how they are similar and different, some of the basics that are so important when it comes to the OSI method and the seven layers that will help us get it all done, and why it is so important to focus on the idea of network security so that our network, and all of the devices and data that are on it, will stay as safe and secure as possible.

There are a lot of things that are going to happen when we spend some time on the security of our network. And you can explore them all in this guidebook. There are so many people who will just sit and use their computers, without having any idea of what is behind them, what makes them work, or even much about the network that is going to drive all of it forward along the way.

And this is something that we are going to work to remedy with this guidebook. The hope is that when you are done, you will have a better understanding of how this is going to work, and what you are able to do with it for your own personal or professional needs.

There are many different parts that we need to learn about when it is time to bring in networking on your computer, and this guidebook aimed to get you started and help you to understand quite a bit about it. When you are ready to dig down under the surface and learn more about how your computer is meant to work and about the network that will drive your computer and all of the other devices that you use on a regular basis, make sure to check out this guidebook to help you get started.

There are plenty of books on this subject on the market, thanks again for choosing this one! Every effort was made to ensure it is full of as much useful information as possible. Please enjoy!

Chapter 1: What is Computer Networking?

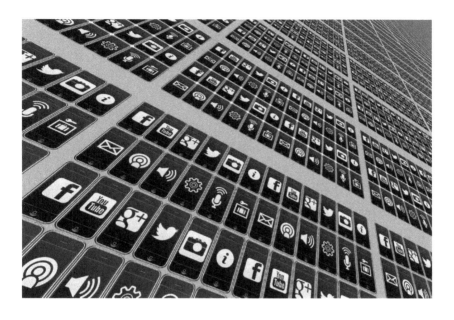

The first thing that we need to take a look at in this guidebook is the basics of computer networking and how important this networking can be to the computers we use. Computer networking is the practice where we are able to interface two or more computing devices so that they can work with one another and share data. These computer networks are built up with a combination of hardware and software, based on what we plan to work with them on.

Keep in mind that we are going to spend our time working on wireless networking and computer networks. These are related to what we can see when we talk about social networking, but they are different topics, so we have to remember this when doing some of the work that we would like. With this in mind, let's take a closer look at some of the things that we are able to do when it comes to computer networking and what all of this is going to mean for our needs.

Computer Network Classifications

First on the list is the different types of classification we can find on our computer networks and a look at the area networks that are important to all of this as well. Computer networks are categorized in a few different ways. One of the approaches is going to define the type of network, and it will do this based on the geographic area that your computer is on at that time. The LANs, or local area networks, are going to span just one single home, school, or even a smaller office building.

Then we can work on a network or something that is a little bit bigger as well. We could handle a WAN instead, which is going to be the wide area networks. These, as we can guess here, are going to span a much bigger area overall, and they will often be found going across cities, states, and other countries. A good example of a large WAN that we can work with is going to be the internet.

Network Design

Computer networks are going to have different approaches to the design that they use.. That means we are going to just focus on the two basic forms of this kind of design including the peer to peer and the client to server option.

To look at these individually, we first need to focus on the client to server networks. These are going to feature some centralized server components that are able to store email, files, web pages, and applications that we can easily get accessed by client computers and some of the other client devices as well.

Then we can move on to looking at the peer to peer networks. These are going to have it set up so that all of the devices are going to support the same functions as well. The client to server networks is going to be the most common when we look towards business arrangements and setups, and then the peer to peer networks are going to be found in personal homes.

Another thing that we need to look at is the network topology. This one can help us define the layout of the network or the structure, and it will take the point of view of data flow to help it get things done. In a bus network, for example, all of the computers are going to share and communicate through one common conduit. But if we are working with a network that is considered a start network, then the data is going to flow through just one centralized device. Some of the most common of these topologies of the network that we are able to focus on will include the mesh, ring, star, and bus.

The Network Protocols

We are going to take a look at these in more depth later on, but we need to spend some time working with these protocols here and seeing what they are going to mean in some of the work that we are doing here. Communication languages that computer devices will use are going to be known as the network protocols. Another way to help us to classify one of these networks is to look at the kinds of protocols that they are able to support.

You may find as you go through the process of learning more about computer networking that many networks can implement more than one type of protocol at the same time, and each network is able to support a specific type of application as well. There are a lot of popular protocols out there that we can work with including TCP/IP. This is a good example of the protocol that we are going to find with some home networks and on the internet as well.

These protocols are important for any network you spend time on. We may realize the importance of these early on when we work with a business computer or system. But even individual systems will need these protocols in place as well.

These protocols are there for a good reason too. They make sure that the information we send online, through email, social media, and even the searches that we do online, are going to make it to the intended recipient in time. We can even choose a good protocol to work with for our intranet within a business, to keep those outside of the business out, and those who belong on the intranet, in and secure.

Choosing the right protocol will help you to accomplish your goals. Everyone, whether they are working with a business computer network or their own personal network, wants to make sure that the data they are sending over will stay safe and secure. Just the thought of someone else reading our messages without permission can make us shiver. These internet and network protocols will make sure that hackers and others stay out.

The Hardware and Software Needed

Another thing that we need to focus on when we are looking at this topic is some of the hardware and the software that we will need to work with on computer networking. There are a lot of good communication devices that are going to have a special purpose when it comes to working on computer networking.

Some of these will include the network cables, access points, and the routers of the network are going to all come together and try to glue and keep the network working well. the operating systems and some of the other applications of software are going to help us to generate the right traffic of the network, and makes it easier for the users to handle some of the useful things that they eed to get done.

Home Computer Networking

We now need to split things up between our home computers and our business computers. Many of us spend a lot of time doing work on our computers when we are not in the office, such as browsing social media, doing searches and more. And if there are more than one option that is available and hooking up to the internet at home, then you already have your own computer network to work with.

While there are a lot of networks that are out there that are built up and even maintained by engineers to keep them working well and to ensure that they are as safe as possible, home networks are a bit different. These are going to belong to the homeowner, and these individuals often do not have much of a technical background at all. Think about how much you personally know about a computer network. You do not have the same knowledge as an engineer, and that can make running your own home network even harder.

The good news here is that there are options that you can work with that will ensure that you set up that network and keep it safe and secure, even when you do not have the technical knowledge that is necessary to get it done. For example, there are a lot of manufacturers who are going to produce hardware for the broadband router that is designed to make the setup of a home network as easy as possible.

A home router is going to be important here because it allows us to have devices working in different rooms, while still sharing the same broadband internet connection in a simple and efficient manner. It can even come into play when people in the home would like to share the same files and even the same printers within the network. And when you pick out the right router to do the job, it is going to ensure that your overall security on the network is improved.

Often, the router that you use in your home will be provided by your Internet Provider. If it is a newer router and has the right security on it, then this router is fine to use. If you would like to personalize it a bit more and make sure that you have a strong firewall and other features on the router, it may be worth your time to search for another router to suit your needs.

You will find that these home networks are going to have an increased capability as the new technology keeps coming up. Just a few years ago, people would commonly set up a home network so that it would be able to connect a few computers, share documents, and even share a printer. This was about all that the router was able to handle at the time.

Today though, things are going to be a lot different. There are many other things that we want to be able to attach to the system, and we want to make sure that we are able to get it all to work in the right manner. For example, today it is more common for households to network smartphones, digital video recorders, game consoles, and more together while still working with their own computers on that network as well.

The good news is that we can work with setting up this network and ensuring that it is going to work the way that we would like as well. Knowing how all of this is going to work, and how to keep the network safe, will ensure that we are able to utilize the network as much as we would like, and makes it so much easier for us to really see some results with using it as much as we would like.

Business Computer Networks

Small and home office environments are going to rely on some of the same technology that we are used to seeing when it comes to the home networks that we talked about before. But sometimes, a business, even if they are smaller, are going to have some additional requirements when it comes to security, data storage, and communication. And this may mean that we need to take the right steps to expand out our network. And sometimes, the methods that we need to use to expand out the network will be varied and different. It often depends on what our goals are overall with this process. And as your business grows and grows, it is going to need to expand even more as well in the process.

While a home network, even with a lot of different devices being hooked up to it, is going to function as one LAN for us to keep things simple, it is common for a business network to have many LANs to it. Companies that have buildings that are in a lot of different locations will utilize what is known as wide-area networking in order to connect all of the branch offices together, even if they are in a wide variety of areas as well.

Though also available and something that homes are able to use on their own network if they would like, voice over IP communication, backup technologies, and network storage are going to be prevalent when we look at the network of a business and some of the extensions that they are working with as well.. Bigger companies can add in a few more things like internal web sites, which are often known as intranets, in order to make it easier and safer for employee business communication to happen.

Other Things to Consider About our Network

There are many other topics that we need to spend our time on when it is time to work with these computer networks. And it is often dependent on what we plan to use the network for, where we plan to use the network, and even how many items and devices we plan to add to the network as well.

The first thing to explore is how networking and the internet work with one another. The popularity of a computer network saw a huge increase with the creation of WWW, or the world wide web, in the 1990s. There are a lot of options that appeared at this time, including peer to peer file sharing system, public web sites, and a lot of other services that are able to run these servers online all around the world.

The internet is going to be so important to a lot of the work that we are doing when it comes to the network that we are working with, as well. Even when a company is going to work with an intranet, we will find that it is important for us to bring on the internet and make this work for some of the needs that we have as well.

This is why we are going to find most of the networks out there hooked to the internet, and why we need to carefully consider how we are will maintain some of the security that we need with our network along the way. It is likely that, no matter what you use your network for, you do not want to allow others to get onto it and steal some of the personal and private information that you choose to store there. That is why there are usually some protocols and other important factors that need to come into play when you connect your network to the internet.

With this in mind, we also need to have a discussion on the difference of wired versus wireless computer networking and then look at how they are similar, how they are different, and some of the benefits that come with both of them.

You will find that many of the same protocols that you will choose to use when it comes to the creation of a new network, including the TCP/IP option, are going to work with both wired and the wireless networks. Networks that rely on the use of the Ethernet cord are going to be found mostly in the past, and it was pretty common to find these in homes, schools, and businesses for many years.

These are good networks to work with. The wired networks were really secure and easy to work with and allowed us to get connected to the internet quickly and efficiently. But they gave the way to the wireless networks because these were more convenient and allowed us to move around and use our devices anywhere that we would like, rather than having to be stuck to the wall to get the work done.

Wi-Fi has emerged in recent years as a preferred option for building new networks for our computers. There are many reasons for this, including ease of use, smartphones, and some of the other wireless gadgets that are out there and are gaining a lot of popularity. While we are still able to find some computers that rely on ethernet cables out there, for the most part, people have switched over to working with the wireless networks, and this is likely where you will build up your own computer network as well.

Keep in mind that if you work with this one that you will want to be careful with some of the work that you are handling. These are not as secure as some of the options that were available in the past, and this can leave you vulnerable to the attacks from hackers along the way. If you are careful about the information that is being sent out, you choose to work with some encryption along the way when you send out messages, and you ensure that there are some tests and checks on the system on a regular basis, you can ensure that even a wireless network can remain safe and secure.

As we can see here, there are a lot of different parts that come together when it is time to handle how things are going to work for you. Whether you are just hooking up your own computer to a network to get online, or you are hoping to combine together two or more computers and make a big network, some of the topics that we are going to find throughout this guidebook can help us to work with this and ensure that it is all up and running in the manner that we need for success.

Chapter 2: The Main Components of a Computer

Now it is time for us look at the actual computer that we plan to work with along the way. We already took some time to look at the computer network and all of the neat things that we should be able to do with that networking. But now it is time for us to understand some of the main parts of the computer, mainly so that we are able to use them a little bit better.

We all know some of the basics that are on the outside of our system. We know the physical parts that come together and allow us to actually work with the network that we are on, or at least with the computer. For example, if I talk about the screen or monitor, the operating system, the keyboard, and the mouse, it is likely that you know exactly which parts I am talking about and can point them out in no time.

We even know a few more. We can name some of the different components that are present when we work with our applications and some of the software that we can find on the computer, and that we plan to use on a regular basis. We can name off the operating system that runs on our computer, even if we don't really understand what all of that means and why it is so important. We can name off a few of the different programs that are found in our computer as well, such as the anti-virus, the writing programs, and more.

In some cases, we may even have a good understanding of some of the networking parts. We may know which Wi-Fi network we are on, and we can remember the username and password that will let us on it anytime that we would like. This is just the basics of what we are going to see when it comes to networking online, and we will talk more about that in a moment, but it does show us an important part and makes networking a bit more real to us.

There are a lot of different parts that are found when we talk about a computer. And even though we are able to list off quite a few of them based on what we use with the computer and how long we have worked with networking there are a few other parts that we are going to spend our time on here. In specific, we are going to look at some of the parts that specifically have to do with the computer networking that we want to focus on in this guidebook.

We could spend our time going over every little component that is going to show up in the computer, but this would take a lot of time and hassle. Instead, we are going to take some time to look at the parts that are the most important, the ones that are going to help us to get right into the process and see some results along the way. For now, we are going to look at the five main components that will help us see some success with the networking that we want to do including the input unit, the arithmetic, and logic unit, the memory or the storage unit, the output unit, and then the control unit. Let's dive in and see what these are going to be able to do for us.

The Input Unit

The first option that we will spend some time looking at is going to be the input unit. This is the unit that is able to transfer all of the information from outside over to the storage or the memory unit. There are a few methods that can be used to see this one happens including:

1. Punch cards that are going to go through a punched card reader and will tell us that information as well.
2. Punched paper tape. This is going to be the kind that is passed through a punched card reader, as well.
3. The last one is a bit more involved. This is going to be a disk or some magnetic tape that is going to be passed through either a disc reader or a magnetic tape reader. A magnetic tape option like this one is going to be similar to what we are used to seeing with a domestic tape recorder. There are going to be two reels and a mechanism for reading and writing all of the data that we would like. You can choose to keep the data that you taped permanently, or you can erase it when you record some data on it. The tape that we see with this option is going to be about half an inch wide and of varying lengths, and the material is usually similar to plastic.

Out of all the options, the magnetic tape is often the best option because it is a lot faster to work with than the punched cards. And like what we see with the punched cards, they are used not only as of the input device, but they can be used to store and then record the output that we want as well.

The magnetic tape is also nice because you can easily make the corrections that you need on them. Additional data can be added to the magnetic tape along the way, which is something that the punched card or the punched paper tape can't do once they have been punched in the first place. But the choice is going to be up to you which of these is going to be the best.

Memory or the Storage Unit

The next component that we need to spend some of our time on is the memory or the storage unit. A series of figures are stored in this unit for us to use later, and then will be released at electronic speed to undergo some calculations along the way. This unit retains temporarily the results of some of the sub-calculations until there is some further processing that goes on as well. this unit is also going to serve the purpose of a store for all of the final results that you get out of those calculations, and then those are going to be passed on to the output unit that we are going to talk about in a bit.

The cost of the computer, and the size of any problem that it is able to handle and take on is going to depend on the capacity that it has for internal storage. The term buffer is one that you have probably heard at some point, and it is going to be used to help with the storage area in a computer, which is going to temporarily hold onto data after the input, at least until it is ready to go out as the output.

Arithmetic or Logic Unit

The third component that we can work with when it comes to our computers is the arithmetic and the logic unit. Required calculations and the logical operations are going to be done when we are in this kind of unit, and they help us take the necessary information out of the memory or the storage unit that we talked about before.

The neat thing about the arithmetic unit is that it comes with the capability of performing all of the ordinary operations that we need with addition, multiplication, subtraction, and division, but it is able to do this at really high speeds. This allows us to get the work done that we would like, without having to wait for the computer to handle it for us.

Then we can look at the logic unit that comes with this as well. The logic unit is going to be the part that we use to make some decisions where the instructions that are provided to our computer require a decision to be made. Such a decision is going to be yes, or no, so make sure to stick with all of that.

Output Unit

The fourth component is the output unit. This is the unit that is responsible for turning out the end product that we are going to work with here. This means that it is the part that will provide us with the result, report, or the final information to be obtained off the computer, including the variances, the economic order quantity of the inventory, and more.

This is going to be the part that we are actually able to see. When all of the information and the steps that happen in the other parts are all done, then it is going to show up on the screen, or in another format, as the output unit. For those who do not know much about the computer world and how all of this sets together, you will be used to just working with the output unit and not some of the other options at all.

The Control Units

While we are in these parts, we also need to take some time to work with what is known as the control unit. This is a unit that will sometimes be known as the program controller, as well. And out of all the other parts that we have talked about, it is one of the most complex options that we to spend some time discussing. It is basically nerve center of this whole process because each of the other units of your computer will work with this one, and the control unit is the part that will supervise all of the other ones as well.

Once you have fed all of the figures that you need to the input unit in the beginning, the control unit will take complete control, and then will handle all of the figures that it seems more like the instructions that the computer program should follow.

One thing that you will notice is that physically, the main component of the computer is going to be the CPU, the central processing unit, which is going to also be known as the central computer. This is an important part when it comes to working with the control unit, so we are going to spend some time taking a look at it here as well. the central processing unit is going to contain some of the following units so we can see more about how this is meant to work as well. It contains the following units:

1. The arithmetic and the logic unit
2. The memory or the storage unit
3. The control unit

Then we also need to take a closer look at some of the hardware and the software that are necessary to make all of this come together. The physical computer system, which is the computer on its own, including the printer, mouse, keyboard, machine, and other related equipment, will all use the electronic data processing system that is known as the hardware. The computer is not able to run and give us the results if it were not able to utilize some of the hardware to help it run.

All of the other materials that we may choose to use when it is time to select, install, and operate our electronic processing system, with the exception of the operating personnel, are going to be the software. This could be things like the operating system that we want to rely on, the applications that we set on it, and even things like Word that help us get work done.

The software is going to be fun because we get a lot of options on what is going to work for us and what we are hoping to gain out of it. The software that we are talking about here is going to include not only the computer programs, which are going to be the steps or the instructions that we are given to the computer, but everything about the Electronic Data Processing System, which is going to be important because it will help the actual machine of the computer, and the reset of the equipment that goes with it, to actually perform their own functions in the process.

Both of these parts are important to some of the work that we would like to do. Software is an integral part of the hardware.

And you will find that one is not going to be able to do the work that it needs without the other one. Putting on the best hardware and the best software that you can make a world of difference in the options that we are going to see and will help you to get the most out of your computer network as well.

As we can see, there are a number of parts that are going to come into play when it is time to work with our computer network. We need to have all of these parts in place to ensure that we are going to see the best results and to help us out when it is time to handle some of the different parts of the network as well. this is going to ensure that the network will do what we are expecting and will not run into problems along the way.

Chapter 3: The Different Types of Computer Networks

The next thing that we need to spend some time on is the kinds of computer networks. A computer network is a group of computers that are going to be linked together in some manner and will enable the computers to communicate with one another, share resources, applications, and even data. A computer network is going to be categorized by the size they come in. The four main types that we are going to take a look at here will include:

1. The Local Area Network or LAN
2. Personal Area Network or PAN
3. Metropolitan Area Network or MAN
4. Wide Area Network or Wan.

Let's take a look at how each of these will work for some of our needs, and what each one is going to mean when it is time to handle some of the work that we would like to do with our own computer networks along the way.

The Local Area Network

The first computer network on our list is going to be the Local Area Network or LAN. This is basically a group of computers that are connected to one another in a small area. This could be in an office or another small building. The connection is not going to be outside of that smaller building, but the network is often going to be a bit bigger than what you would find in your own home.

The LAN is a good network option to work with in order to connect two or more computers through a communication medium, including the coaxial cable, twisted pair, and more. And when we look at the price of this one, we will notice that it is going to cost less than some of the other options we will talk about in a moment. This is because we are able to build it with some of the more inexpensive hardware options including ethernet cables, network adapters, and hubs to name a few.

When we are working with the LAN, the data is transferred at a rate that is much faster than what we are able to find with some of the other options out there. This can be good if you need to be able to share and access the data in an efficient manner that doesn't take that long. And many offices and businesses are going to enjoy that the Local Area Network is going to provide them with some more security, so they don't have to worry about how safe their data is during the process.

The Personal Area Network

Next on this list is the Personal Area Network or the PAN. This is a network that is arranged for just one individual person to work with it. Usually, the range of this is within ten meters or less. The PAN is a good network to use when you would like to connect a computer or another device that you would like to use just for your own personal needs, rather than needing it for business purposes or something else.

Thomas Zimmerman was actually one of the first research scientists who were able to bring the idea of this PAN to the public, and it was able to help us to really use our own computers and devices at home, even if we did not need to share a lot of information. Personal computer devices that are used to develop and work with this network are going to include options like a laptop, play stations, media players, and mobile phones.

Another thing to consider is that there are going to be two main types of personal area networks that we are able to spend some time on. These include the wireless personal area network and the wired personal area network.

You will find that the wireless network option is developed with some simple wireless technologies. This is going to include Bluetooth and Wi-Fi options so that we can move around and enjoy the use of the network without being stuck to the wall. The range with this is often pretty low, but this is beneficial to the security that you will require to keep your network safe and secure. And then there is going to be the wired personal area network. This is going to be the network that we are able to create with the help of a USB.

To help us see a bit more about how this personal area network is going to get set up, we need to first take a look at some of the examples of a personal area network. These will include:

1. The Body Area Network. This is the network that is able to move around with the person who is using it. For example, the network that you use on your mobile device is an example of one of these. You can still use it to make your own connection and then create a connection over to one of the other devices that you would like.
2. Offline network: This kind of network is created inside of our home, so it is known as the home network. This is designed in a manner to help integrate the devices, such as television, computer, and even printers. The thing to remember with these though, is that they are not going to be connected back to the internet as your computer will.
3. Small home office: This kind is used to help us connect a lot of different devices to the internet and to a corporate network. To keep these connections as safe and secure as possible, there is usually a VPN attached to it as well.

Metropolitan Area Network

While you may have heard about the other two options that we talked about in the past, the MAN is a bit new, and we need to spend some time taking a look at how this works. The metropolitan area network is going to be an option that is able to cover a larger area by interconnecting a different LAN and basically creating and using a larger network.

Many government agencies use MAN to help them have a better connection and share information better with both private industries and citizens. In MAN, various LANs are connected to one another with the help of the telephone exchange line. There are a number of protocols that need to be used in order to help us get this done.

In order to ensure that the right people are getting the information that they need through this process, the MAN comes with a much higher range than we see with the Local Area Network that we talked about before. This ensures that all of the computers that are on that network are able to get the data and the information that they need.

There are a few different reasons that we are able to use this MAN to help us get things done. Some of the uses of the Metropolitan Area Network include:

1. This is a good system to work with to ensure that the banks in a big city are able to communicate with one another.
2. It can be used to help when we are working with an airline reservation.

3. Many times colleges in a big city will use this to keep all of the computers in that network working with one another.
4. It is also a good way for the military to work with and keep their communications working well between one another without someone else being able to get on if the right security is in place.

If we are working with one of these networks, it is so important that the right kind of security is put in place before it is even started. These networks are so large, and that is going to bring out a lot of the vulnerabilities that we need to be careful about as well. When the right security is in place, though, this is going to help ensure that the network will stay safe, even if it does need to grow to be pretty large.

Wide Area Network

Another thing that we need to look at is the fourth type of network, known as the Wide Area Network or WAN. This is a type of network that can be expanded out over a larger geographical area, including over states and even over whole countries if we would like. As you can imagine, the WAN is going quite a bit larger than we see with LAN, or even with the other options that we have talked about in this chapter so far.

The Wide Area Network is also unique in that it does not have to keep itself limited to just one single location. It is able to span over a really large area geographically through the fiber optic cable, telephone lines, or satellite links as well. and we will find that the internet is going to be one of the biggest WAN examples throughout the whole world for us to rely on.

There are many times when the WAN is going to be used, but we will often find it in a few industries who will use it quite a bit. These industries are going to include education, government, and even business. These kinds of networks are going to help them to figure out how to communicate with their customers, and even with other branches that they own, even if all of these are in a completely different location.

There are a lot of examples of this kind of network as well. Some of the most common examples of the Wide Area Network that we are going to see will include:

1. Mobile Broadband: You will find that for most smartphones, the 4G network is something that they are going to rely on. This is often used across a whole region or a country.
2. Last mile: A telecom company is used to providing the right internet services to their customers. And they often have to do this across hundreds of cities and more. This

happens when the company connects the home of the customer with the fiber that is needed.
3. Private network: A bank is able to provide its own private network, for example, that is able to connect the 44 offices. This network is going to be made with the help of the telephone leased lines that are provided through the telecom company.

Now, there are a number of benefits that come up when we want to work with the WAN. For example, it takes up a really large geographical area and can ensure that we are able to find the parts that we need in a short amount of time. Suppose if you are working with a branch of your office, that is another city. You would be able to use something like the WAN in order to connect with them. The internet is the perfect place for us to have a direct line over to this other office so that we can connect and communicate as needed.

Another benefit is that it provides us with some centralized data. In the case of the WAN network, the data that we want to work is centralized. Therefore, we are not going to need to worry about working with backup servers, files, or emails in order to help us keep this information safe and to avoid losing the information either. And along this idea, we will find that it provides us with some of the updated files that we need.

Software companies work on the live server for example. This means that the programmers will end up with updated files in just seconds.

The WAN is also able to help make it easier to exchange some of the messages that we need. And these messages are going to be transmitted quickly. The web applications that we use on a regular basis, like Whatsapp, Skype, and Facebook make it possible to communicate almost instantly with your friends and family so you will not need to worry about delays or all of this taking too long either.

Some companies like to work with this WAN because it makes it easier to share the software and resources that they would like. In this kind of network, it is easier to share all of the hardware and software that we would like, including the RAM and the hard drive. And since it is made to work with global businesses, your company is not limited by geographic boundaries any longer. You can handle all of this in an easy to handle manner.

The final benefit that is important here is the high bandwidth. If we use some of these leased lines on our company, then this is going to provide us with a high bandwidth as well. this is a good option because it will increase the rate of data transfer, which is then going to increase how productive the company can be as well.

However, while there are a lot of benefits that we can see when it comes to working with this Wide Area Network, it is also important to realize that there are some disadvantages along the way as well. Some of these include:

1. Issues with security. This kind of network is going to have more issues with security than some of the other networks. This is mostly because it is so large, and all of the technologies are combined together to create some security problems. It is important to be aware of these if you would like to keep your information safe and secure when using the WAN option.
2. You have to work with software for an antivirus or a firewall: The data that we are talking about here is transferred over the internet for you and the other person to communicate. But if a hacker is able to get onto the network, they can change or steal the information instead. This is why a firewall is so important. And it is also possible that a hacker can add in a virus to ruin your documents and your system, so having an anti-virus in place is important.
3. The costs to set up are high: The cost to install the WAN is higher because you have to connect a lot of different

computers, and you have to purchase a lot of things like switches and routers.
4. Problems with troubleshooting: The WAN is will cover a big area, so you will find that when an error shows up, it is going to be hard to figure out where it is in the first place.

InternetWork

Another thing that we need to spend some time on while we are here is known as internetwork. This is going to be defined as two or more computer network WAN or LANS, or a computer network segments that will be connected using devices, and they will be confused by a local addressing scheme. This process is going to be great because it is known as internetworking.

Another thing that we are able to look at in this process of internetworking is that it is going to be when we are able to interconnect between government, industrial commercial, private, and public computer networks. This helps us to get a bunch of parts to combine together well and give us the results that we are looking for. And to get internetworking to behave properly, we have to make sure that we are working with an internet protocol. The reference model that is used with this one is going to be OSI, or Open System Interconnection. We will spend some time looking at this in more details later on in this guidebook.

The first type of an internetwork that we need to look at is the extranet. This is a type of communication network that is based on several internet protocols like the Internet Protocol and Transmission Control Protocol. It is going to be used when you would like to share information. The access to this is going to have a lot of restrictions, and only those who have been given the proper credentials for logging in will be able to access this network at all.

The extranet is the lowest level of the internetworking that we are talking about here. It is categorized in a variety of manners, including the MAN, WAN, or some other computer network. An extranet is not going to have just one LAN because it has to be able to connect with, at a minimum, one other external network, and sometimes it is able to hook up to more than one as well. This means that we need to have more than one LAN present in this kind of system.

The second type of internetwork that we will focus on is the intranet. This is a private type of network that is going to work on the same kinds of internet protocols that we find with the other option. This is a network that belongs to a specific organization, and only the employees of that company, or members who are approved for another reason, are going to be allowed to gain any kind of access to this system or network.

The main goal that we see when we look at the intranet is to share the information and the resources that are needed among the employees of a particular company. An intranet is going to provide the facility to work in groups, and even for teleconferences as well.

There are a number of advantages that we will see when it comes to working with the intranet. Some of these will include:

1. Communication: This is going to provide us with a cheap and easy form of communication. An employee with the company can communicate with another employee with the help of chat and email.
2. It saves a lot of time: Information that we find on the intranet is going to be shared in real time, which saves us a lot of time in the process.
3. Collaboration: This is actually one of the biggest advantages that we see when it is time to work with the intranet. The information that we distribute between the employees of a company is easy to do, and only the users who are authorized will be able to access it at all.

4. Platform independency: It is an architecture that is neutral because the computer is able to connect to another device, even if the architecture is a bit different.
5. Cost effective: People will be able to see the data and all of the documents that they need by using the browser and then distributing the duplicate copies over this intranet source. This is going to help keep down the costs that the company is experiencing.

The different types of networks are going to be so important to some of the work that you are able to do in this process. Understanding when each one is going to be used, and what they all mean, is part of the process as well. we can see why they go by different names and how much area they are going to be able to help control. And we understand that the larger these networks get, the harder it is to use them because they become more complicated, and we need to worry about that as well.

Chapter 4: What is the Virtual Memory?

Another topic we need to spend some time talking about when it is time to work with our computer networking is the virtual memory. This is one of the common parts that you will find when it is time to work on an operating system or a desktop computer. In fact, it has become so common because it provides us with a really big benefit, without really costing us all that much in the process.

Most computers come with some amount of RAM, or random-access memory, on them. This is usually something like 64 or 128 megabytes of RAM that we are able to use. And this is specifically set aside to work for the central processing unit or CPU. While this is a good deal of space, we have to keep in mind how much it takes to run some of the other parts of our computer, and this amount of RAM is often not going to be enough to run all of the programs that most people would like to work with at once.

Let's take a look at an example of how this will work in computer networking. If you load up the operating system of Windows, and then an e-mail program, a word process, and a web browser, all of this is hitting the RAM at the same time. When this happens, the 64 megabytes is not enough to hold it all. If there isn't something like a virtual memory, your computer would have to say something like, "Sorry, you are not able to load any more applications. Please close an application to load a new one."

This is not something that most of us would like to hear when we are doing some of our own work along the way, and that is why the virtual memory is set up to come in and help us out here. With this virtual memory, the computer is able to take a look around for some of the areas of RAM that have not really been used as much recently and then will copy them over to the hard disk. This is going to free up some of the space in the RAM so that we can load up that other application that we would like.

Because this is going to be something that happens in an automatic manner, without you doing any of the work or setting it up to do it, then you will often not be noticeable that it is happening. And often, it is going to make it feel like our computer is going to have an unlimited amount of RAM space, even though it has a limited amount. Because the space of the hard-disk is going to be so much cheaper than we are able to find with the chips of the RAM, the virtual memory is a great way to handle the demands that we have on our system without upping the costs.

The area that is responsible for storing the RAM is found on our hard disk, and it is called a page file. This is responsible for holding onto the pages of RAM and placing it on the hard disk, and then the operating system will move the data back and forth between the RAM and the page file. When we are working on the Windows machine, though, the pages are going to look a bit different and will come with the extension of .swp.

Of course, we have to keep in mind that the read and the write speed of a hard drive are going to be a lot slower than what the RAM is able to do, and the technology that we have in our hard drive may not be geared towards accessing all of those smaller pieces of data at a time. If your system starts to rely too much on the virtual memory, even if it is a good thing, then you are going to notice that the performance of your computer will start to slow down and drop along the way.

The key to making sure that your computer is able to handle all of the work that you are throwing at it, and that it can able to handle some of this work and get all of the parts done in the manner that you would like, is to have plenty of RAM to handle everything that you plan to work with at the same time on a regular basis.

Then, the only time that you will feel that the visual memory starts to be a bit slower in all of this is going to be when there is a slight pause that will occur when it is time to change up some of the tasks that you are doing. When you start out with enough RAM to handle some of your needs, then the virtual memory is able to do the work that you would like, and it will work perfectly.

But if you do not have enough RAM, either because you assume that the virtual memory will be able to handle some of the work for you, or you just don't know how much RAM you are using realistically, then the operating system is constantly spending its time swapping information back and forth between your hard disk and the RAM. This is a process that is known as thrashing, and it is going to get really annoying since your computer will start to feel like it is going really slowly in the process.

The virtual memory that you are able to work with is there to ensure that your computer is going to work in the manner that you would like. The goal is to have enough RAM on the computer so that it is able to handle all of the tasks that you would like to get done. But sometimes the RAM is not able to handle it all, and that is where the idea of the virtual memory is going to come in and make life a little bit easier for you as well.

Chapter 5: A Closer Look at the Networking Infrastructure

Another topic that we need to spend some time on here is the network infrastructure. This is basically an interconnected group of computer systems that are linked together with the help of the various parts of the telecommunications architecture. To make this a bit more specific, this is going to refer to the organization of its various parts and their configurations. And this is seen as true whether we are talking about the individual computers that are networked or other parts like the wireless access points, cables, routers, switches, network protocols backbones, and network access methodologies.

These infrastructures can be found either as closed or open. We could find something like the openness that is with the Internet, or we can find the options that are closed off and private and are something like the intranet. They can also work over a wireless or a wired network, and sometimes they will be a combination of both.

The most basic form of this kind of infrastructure will consist of one computer and sometimes will have a handful. These will also come with a network or some kind of connection over to the Internet, and then a hub that is going to help link all of those computers over to the network connection, as well as to one another on that same network. Keep in mind here that the hub is merely there to link the computers, but it is not there to limit any of the low of data that comes into or out of any one of the systems.

Now, sometimes you want to control or limit the access between the systems and regulate the information flow. To do this, you will need to put in the switch and have it replace the hub. Then you can use these to help define how the system is going to communicate with the other parts based on your own needs to allow the network that is created by these systems to have the ability to communicate with the others, through that network connection, you will need to work with a router, which is going to bridge the networks and will basically provide us with a common language to help with the exchange of data. For this to work though, we need to follow the rules of each network to share the data.

When there is more than one computer that is found in the same household, and they are sharing the same connection to the internet, this is a good example of what we will see with this kind of network infrastructure as well. This is true whether those computers are set up to share information with one another or if they are just sharing the internet connection.

The internet on its own is another example of the infrastructure that we are talking about here, but it is a bit more advanced and includes a good network of items as well. The Internet is basically where the individual systems are able to gain some access to the global network that houses information on all the various systems and will allow this access with web standards and protocols, most commonly framed as URLs or the addresses that we type in when we want to do a search on something.

Office intranets are similar to what we see with the global internet, but they operate more on a closed network infrastructure compared to the others that we are talking about. This means that only those who are within the network, and the company, will be able to access the network at all. This could consist of a central data store, one or more computers that are known as the servers, and even an ethernet cabling, routers, switches, and wireless access points. And of course, there are some individual computers that are going to be there to access that central store of data.

The individual computers are set up to work by connecting to the network through either wireless access or cabling. The routers and the switches are then going to determine what level of access the computers are going to have at that time. They can also act as the directors of traffic, helping to point them over to the central data store on the servers. As the individual computers send or receive data, the routers are going to ensure that it reaches the right place at the right time. This is a great way to make sure that we get the right information to the right place as much as possible.

Another thing that we need to consider when we are working with this is network security. This is actually a huge concern when we are building up a new infrastructure for our network. Most architectures are going to work with some firewalls and routers and can add in some software that helps us to fine-tune the control of who is able to access the data. There can be some other parts that show up as well, including data packet monitoring to see what is coming in and out of the network, and some protocols that have to be defined pretty strictly.

Security can also be controlled by adjusting some of the network sharing properties on individual systems, which is going to limit the folders and the files that can be seen by others who are on that network at the same time. This gives the owner of that whole process all of the control that they would like to keep it all running as smoothly as possible.

Keep in mind here that the network infrastructure is going to be the resources of software and hardware that an entire network is able to work with. This can include things like network connectivity, operations and management, and the communication of that network. It is going to be able to provide us with a good path of communication and ensures that the right services can happen between processes, users, services, applications, and any of the external networks that we want to interact with, including the internet.

This network infrastructure is typically found alongside the IT infrastructure that is found in some of the enterprise IT environments. The entire network infrastructure is a system that is interconnected, and you are able to use it to help with either the internal communications, the external communications, or both.

We can take a look at some of the parts that are going to show up in one of these network infrastructures. Remember that each one is going to be a bit different compared to each other, based on what information we are trying to look for, and what is going to be important to us at the time. Some of the parts that we can look for includes;

1. The networking hardware
    a. Cables
    b. Wireless routers
    c. Switches
    d. Routers
    e. LAN cards
2. Networking software
    a. Network security applications
    b. Firewall
    c. Operating system
    d. Network management and operations
3. The network services
    a. IP addressing
    b. Wireless protocols
    c. Satellite
    d. DSL

e. T-1 Line

The network infrastructure is important to what we need to do when it comes to handling some of the work that we are doing in computer networking. Being able to handle this part and understanding how it works and what parts come with it can make a big difference in the work that you are able to do with networking as well.

Chapter 6: Understanding the Protocols of a Network

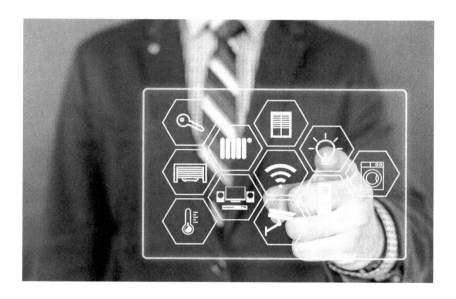

The different types of network protocols that we are able to work with will be very important when it comes to working with computer networking along the way. We have already mentioned a few of these along our journey, even without explaining them, so this can give us a good idea of why they are so important and the fact that we do need to learn more about them and what we are able to do with them.

To start out our journey here, we need to take a look at what a network protocol is all about. These network protocols are responsible for defining the rules and some of the conventions that are needed so that the devices of the network are able to communicate with one another. These protocols on the network will include some of the mechanisms for devices to properly identify and make connections with one another. There will even be some of the formatting rules that will help specify how the data is packaged into sent and received kinds of messages.

These protocols can take it even further than that if we would like. For example, there are some of these that are able to support acknowledgment about the message, and the compression of data. Both of these are designed for reliable and for some high-performance communication on that network if you would like to take it this far.

Some of the modern protocols that we see when we work with computer networking will use something that is known as a packet switching technique in order to help us send and then receive messages that are in the form of packets. These packets are basically just messages that are divided into pieces that are going to be collected and reassembled at their destination. There are hundreds of these protocols for the computer network that are in use and have been developed. You have to pick and choose the one that we like the best based on the purposes and the environments that we are working in, and which one is going to fit into our needs.

Internet Protocols

The first option that we will work with is the Internet Protocol or IP. This is a set of related, and most widely used out of all of them, network protocols. There are actually quite a few that fit into this family and that we can use for some of our own needs along the way, as well.

Outside of the Internet Protocol, we will find that some of the protocols that are higher-level, like the FTP, HTTP, UDP, and TCP, are going to be able to integrate with the IP to provide us with some of the additional capabilities that we would like to work with. In addition, some of the lower-level types of these protocols are going to be able to coexist with the IP as well, including the ICMP and ARP.

In general, some of the protocols that are higher-level in this IP family are able to interact in a more close manner with a lot of different applications, including the web browsers, while some of the lower-level protocols are able to work with things like the hardware on a computer, including our network adapters.

Wireless Network Protocols

In addition to some of the internet protocols that we spent some time talking about before, we are also going to need to spend some time talking about the Wireless Network Protocols.

Thanks to some of the options like LTE, Bluetooth, and Wi-Fi, we will find these wireless networks pretty much all around us all of the time.

These network protocols are a bit different than IP, but they are designed for us to use on a wireless network. And this means that they need to be able to handle a few of the things that you would need, including supporting roaming mobile devices and then deal with issues such as variable data rates and anything that will come with network security.

One thing that we need to work with here is the security. Since you are working with the wireless network, security can be an issue. These will work because they can hook up with the network adapters on your own computer, but you need to make sure that it is only hooked up with your needs, rather than with another computer or device that should not be there. This is a unique balancing act that we need to be careful about along the way, as well.

This is why we need to make sure that this kind of protocol is set up in the right manner. There are a few different protocols that we are able to work with when it comes to our wireless networks, but we have to go with the ones that are going to help us to get things done, and maintain some of the security that we are looking for so that our personal information and more will stay safe and secure.

Network Routing Protocols

Next on the list is the network routing protocols. These routing protocols are going to be pretty special in that they are designed for specific use with the network routers that are online. A routing protocol is able to go through and identify some of the other routers that are present, can help us to manage some of the pathways that are out there, which are known as routes between the destinations and sources of network messages, and will make some of the dynamic routing decisions. Some of the most common routing protocols that we are able to work with on our computer network will include BGP, OSPF, and EIGRP.

Other Protocols to Consider:

There are a few other types of protocols that we are able to work with. The first here is the Internet Protocol Suite. This is a set of protocols on communication protocols that are going to be implemented on the protocol stack where the internet is going to run. This suite is known as the TCP/IP protocol suite because this is based on these two types of protocols, as well. We are able to describe this kind of suite similar to the analogy that is found in the OSI model, but there are a few big differences that we are able to focus on, and not all of the layers are going to line up and correspond in the manner that we would like.

Another option to work with is the protocol stack. This is the complete set of protocol layers that are able to work together and will provide us with all of the networking capabilities that we need. If you would like to do some work with networking, then you will want to work with this special protocol stack.

Next on the list is going to be the TCP or the Transmission Control Protocol. This is one of the core protocols that we are able to use out of the Internet Protocol that we talked about before. It is one that originated in the network implementation and it was meant to help complement the Internet Protocol as well. Because of these two working so closely with one another, we will find that the suite is often called TCP/IP. These two work closely together to help us keep the network safe, and the two protocols will often be found together.

As you work with some of the protocols that are important here, you will find that TCP is going to provide us with a reliable delivery of a stream of octets over an IP network. Error-checking and ordering are going to be some of the main components that we will be able to see with TCP, but it is possible that there will be other uses of this as well. all of the big internet applications that we will want to use, including file transfer, email, and the World Wide Web, are going to rely on the TCP option.

We can also take a look at the part that is known as Internet Protocol. This is the principal protocol that we can work within the Internet Protocol Suite and it is a great option to work with when it is time to relay our data across the network. The routing function that comes with this is going to be the main establishment that we will find with the internet.

In a historical manner, it was going to be the connectionless datagram service in the original TCP, the other being the connection oriented protocol. This is why this is the suite that is going to be known as the TCP/IP that we want to work with.

Some of the different options that we are able to work with when it comes to the protocols that we would like to handle. Some of these are going to include:

1. HTTP: This one is the Hypertext Transfer Protocol. This is part of the foundation that we are able to find when it comes to data communication when we work with the world wide web. This to be the hypertext that is going to be more structured, and it is going to use some hyperlinks between the nodes that contain some texts. The HTTP is an application protocol for distributed and collaborative hypermedia information system. The port that we see

with HTTP is 80, and it is going to have a secured port of 443 if you would like to work with this one.
2. FTP: This is the File Transfer Protocol. This is one of the most common protocols that we are able to work with when it comes to file transferring on the Internet, and it is within the private network that you would like to work with. The default port that we are able to work with when it comes to FTP is going to be 20 or 21.
3. SSH: This is the Secured Shell. This is one of the primary methods that we are able to use in order to help us manage some of the devices on our network in a secure manner, even when we are on a command level. It is usually used as one of the alternatives that we are able to work with from Telnet, and it is not going to support a connection that is secure. The default port that we see with SSH is 22.
4. Telnet: The next option is the Telnet option. This is one of the primary methods that we can use when it is time to manage the devices on our network at a command level. Unlike what we see with SSH, Telnet is not going to provide us with a secure connection, but the unsecured connection is basic and easy to use. The default port that we are able to use with Telnet is going to be 23.
5. SMTP: This is stands for Simple Mail Transfer Protocol. It is used for two main functions. To start, it is used to

help us transfer emails from the source to the destination between two mail servers. And it can also be used in order to transfer email from end users to a mail system. The default port that we are able to use for this one is 25 or the unsecured, and then the secure port, which is not standard, will be 465.

6. DNS: This is going to be known as Domain Name System. DNS is going to be used in order to convert the domain name to the IP address. There are going to be root servers, authoritative servers, ad TLDs. The default port that we are going to find with this one will be 53.
7. POP 3 This one is for Post Office Protocol version 3. This is one of the two main protocols that are used in order to help us retrieve mail from the internet. It is simple because it makes it easier for the client to retrieve complete content when they need it from the server mailbox and then can delete some of the contents from the server. The default port that comes with this one is going to be 110 and then the secured option is going to be 995.
8. IMAP: This one is going to stand for Internet Message Access Protocol. This is another one of the main protocols that is used to retrieve mail from the server. This is not a protocol that will delete the content from the mailbox of the server as well. the default port of this is going to be 143 and the secured port is going to be 993.

9. SNMP: This is the Simple Network Management Protocol. This is used to help us manage the networks we have. It is going to have the ability to configure, monitor, and even control the devices of the network. This is going to be configured on network devices to notify a central server when the specific action is occurring. The default port that we are going to work with here are 161 and 162.
10. HTTPS: This is for the Hypertext Transfer Protocol over the SSL/TLS: This is going to be used along with the HTTP in order to provide us with some of the same services. But you will get a more secure option with this thanks to the TLS and SSL. The default port that you are able to work with when it comes to the HTTPS will be 443.

How These Network Protocols are Implemented

Now that we know a bit more about these network protocols and what we are able to do with them, it is time to take it a bit further and see how these protocols are going to be implemented along the way as well. Some of the modern operating systems that you can choose to work with are going to come with services of the software that are built-in, and that will be able to implement some of the supports that are needed for all of these protocols on the network.

Applications, including your web browser, contain some software libraries that are responsible for supporting the high-level protocols that will help that function to do the work that it should for some of the lower-level routing and TCP/IP protocols, the support is going to be implemented in the direct hardware (which could be something like the silicon shipsets) to ensure that the performance is going to see some improvement in the process.

Each packet that is transmitted and received over our network contains some binary data. This means that we rely on ones and zeros in order to encode the contents of each message that we have. This adds another level to the work that we are doing, but will ensure that we are able to handle the work, and can add in some of the power that these networks need.

Most of the protocols that we want to work with will add in a small kind of header right to the beginning of the packet that we are working with. This helps us to store a bit of important information, such as information concerning the sender of the message, and the destination where that message is supposed to come.

Depending on the protocol that you choose for your project and your network, it is possible that it is going to add in a new footer to the end of it. Each network protocol is able to identify messages of its own kind, and then it will work to process both these footers, and the headers, as part of the data that moves between some of the devices that we are working with.

Another thing to keep in mind is that a group of network protocols that are able to work together at a lower and a higher level will be known as a protocol family. Students who are learning more about networking are going to spend some time on the OSI model, which we will talk about in a bit, to help them organize these network protocol families into specific layers so that they can learn and teach how to work with these as well.

As we can see, there are a lot of different protocols out there, and each one is going to come with a few of the different rules and more that we are able to work with as well. Make sure to figure out which protocol is going to be the best for some of your needs, and work from there in order to ensure that you are getting the right rules and more that we would like to work with.

Chapter 7: How to Handle IP Addressing

When it comes to working with some of the computer networking protocols and processes that we have discussed so far in this guidebook, you need to spend some time working with the IP addresses. These are going to help identify your computer and will allow it to have an easy option when it is time to handle any communication on the network, and even the communication that is going to happen with the outside sources and computers as well. With this in mind, we need to spend some time working with how to handle these IP addressing and how important this is.

Every device that is connected over to a network, including cameras, printers, tablet, and computers, need to come with a unique identifier so that the other devices know how to interact with and reach it. In the world of TCP/IP networking, this is known as the IP or the Internet Protocol address.

If you have spent any time working with a computer for any amount of time, then you have spent some time working with these IP addresses, which are a kind of numerical sequence that will look like 192.168.0.15 or something similar to this. Most of the time, we are not going to deal with this in a direct manner, because the network we are on, along with the device that we are using, will often just be set up in order to follow instructions about what numbers to put where. But if you have ever wanted to learn more about how to work with this, and dive into this a bit more, about what these numbers really mean and how they work, then this is going to be the right chapter for you to spend some time with.

Having a good understanding of how these IIP addresses work is so important if you would ever like to do some troubleshooting about why the network you are on is not working I the proper manner, or why you can't get one of your devices to connect in the way that it should. And, if you end up ever needing to set up something that is more advanced than some of the basic computing hat we are looking at, such as a media server or a game server so that you can connect with some of your friends online, then you have to understand a bit more about this IP addressing.

In addition, you will find that there is a lot of information that is found in these addresses, and knowing what yours is, and how they work, is going to make a big difference in the amount of success that you are going to be able to see overall. Setting the IP address up and knowing how they connect you to the internet, and how these allow for some communication between your devices, can make a big difference in the amount of success that we see.

Note, we are going to spend some time covering the basics of this IP addressing that is in this chapter, the stuff that you are going to be able to use with an IP address but who haven't put much thought in it, and they want to know some more. This will not go into some of the more advanced options, but it will help us to go through and learn more about how this is going to work for our needs.

What is the IP Address?

The first thing that we need to take a look at is what the IP address is all about. This is a unique address that is able to identify the device found on your network. It is likely that at some point, even if you did not recognize them before, you saw these addresses show up. They will be different based on the device that you are working with, but they will look something like 192.168.1.34.

An IP address is always going to be set up into four sets of numbers, similar to the option that we saw above. Each of the numbers is going to range from 0 to 255. So, the full IP address range is going to be from 0.0.0.0 all the way to 255.255.255.255.

The reason that each of these numbers is only able to reach up to the number of 255 is that each of these is really an eight digit number that is binary, and sometimes this is known as an octet. In these octets, the numbers zero out be 00000000 while the number 255 would be 11111111, the maximum number that our octet is able to reach. So, the numbers would be set up in order to handle the idea of binary as well.

Computers work with this binary format, but for the most part, we will find that it is easier for us to work with the decimal format and understand it a bit better. But when we know that these IP addresses are actually supposed to be binary numbers, even if it doesn't look like it, because it helps us to learn a bit more about how these IP addresses are going to work in that specific way. The good news is that while it is possible to spend some time on it, we are not going to throw a ton of math or binary at you. We are just going to get the basics so that you know how to work with this.

Another thing that we need to explore when it comes to the IP address of the device is that it is going to come in two separate parts. These include the Network ID and the Host ID.

The Network ID is the part of the IP address that starts from the let that will be able to identify the specific network on which you will find the device. On a typical home network that we would spend our time on, where the device is going to come with an IP address of 192.168.1.34, you will find that the first part, the 192.168.1 is the network ID. It is custom to fill in any of the missing parts with a zero so that you may find that when you are asked for the network ID of this, you will end up with the results of 192.168.1.0.

Then there is the host ID. This is a part of the IP address that is not going to fit into the network ID. It is there to identify some of the specific device (when we are in the world of the TCP/IP protocol, we call these devices the hosts) that are found on that network. Continuing the example from above, the IP address is going to have a host that is 34 for this one.

If you take a look at these and try to see what ones are on your home network, then you may find that there are several devices and they will come with their own unique IP addresses. This is because each of the devices will come with their own unique IP address, even though they are on the same network.

To help us get a better understanding of how this is going to work, we can look at it as an analogy. This is a process that is similar to how the street addresses in your town are going to be. We could take the address of 2013 Paradise Street and work with it here. The street name is similar to what we see with the network ID. And then the number of the house is like the host ID.

Within a city, you are not going to find two streets with the same name because this is would make it really confusing when you are trying to find what you would like. And this is why you will not find any two network IDs that are on the same network with the name the same either. On a particular street, each of the house numbers will be unique, just like what we see with the host IDs within a particular network ID will be unique as well.

The Subnet Mask

So, we now need to look at how our device is going to determine which part of the IP address is seen as the network ID and which one is the host ID? For this, the computer is able to work with a second number, which is always seen in association with the IP address and this is known as the subnet mask.

On even the most simple of these networks, like those that you use for your small business or your home, you are to see that there is a subnet mask that is like 255.255.255.0, where all four numbers are going to fall under the 0 or 255 part. The position of the changes that happen between these indicates the division between the network and the host ID. The 255's are important because they will mask out the network ID from the equation that we are working with.

Note, we are talking about some of the subnet masks here, and they are known as the default subnet masks. Things get more complicated than this on the bigger networks. People are often going to use some of their own custom subnet masks (where the position of the break between zeros and ones will shift within the octet), to help them create more than one of these subnets on the same network. This is a bit more than what we talk about in this chapter, but it is still an important part of the process and will help us to see the results that we would like in no time.

The Default Gateway Address

The next thing that we need to look at is the default gateway address. In addition to the IP address that we have been talking about so far, and the subnet mask that is associated with it, you will also see that there is the default gateway address that is listed along with the IP addressing information. Depending on the kind of platform that you are using, this address might be called something different.

Sometimes we see this go by another name like gateway, default route, router address, and router. These are all considered the same thing. It is more of the default IP address to which the device is able to send out some of the data on the network when that data is supposed to head over to an external network (one that has a different network ID), than the one that the device is on.

We can find a nice simple option of this in the typical home network. If you are working with your own home network and there is more than one device on it, then it is likely that you are working with a router that is then connected over to the internet through a modem. That router is sometimes a separate device, and other times it is a big part of the modern or router combo unit that your internet provider has been able to give to you.

The router is important to some of the networking that we use because it sits between the devices and the computers that are on your network, and the more public-facing devices that are on the internet. The router is also responsible for routing, also known as passing, the traffic back and forth and making sure that it is safe and will be able to reach the right device.

Let's say that you get started and fire up your browser and then head to a website of your choice. Then the computer is going to send that request to the IP address of the website that you want. Since the server is on the internet, rather than found on your home network, that means that the traffic is sent from your PC over to the gateway or to your router, and then the router is going to forward this request on to the server of that website.

The server, when it gets the information, is going to send the right kind of information back to your router. The router is then able to move the information back to the device on your network that requested it in the first place. And if it all works in the manner that it should, then you will see that the requested site is going to show up on your browser.

The DNS Servers

And now, we need to take a look at one final piece of information in order to really make this work and to ensure the IP address is going to behave in the manner that you would like. One thing that you are going to find alongside a device's IP address, the default gateway address, and the subnet mask that we have been talking about is the address of one or two DNS servers that are the default for this one.

Remember that for most people who are working with the network, the name is going to be a lot easier to work with and understand compared to the numerical address. We are more likely to type in the name of the website, which is usually in words or phrases, rather than typing in the IP address, which is going to be easier for all of us.

The DNS is similar to what we see with a phone book because it is going to help us look up some human-readable things like the names of websites and converting these over to the IP address that you need. This is done by the DNS by storing all of that information on a system that contains the linked DNS servers throughout the internet. Your devices have to know the addresses of the DNS servers so that they know where to send the queries.

When you are working with a typical home or small network, the DNS server IP addresses are often the same thing that we are going to see with that default gateway address that we talked about before. The device is going to send in these queries of DNS to our routers, which are then able to forward on these queries to whichever NS servers the router is configured to use at this time.

When we are looking at the default, these are usually the DNS servers that the ISP provides, but you are able to change these up and choose another option for the DNS servers if you would like. If you do wish to do this, you may find that you are able to work with better success using DNS servers provided by some of the third-parties that are out there, including OpenDNS, and Google if you would like.

Chapter 8: A Look at the CISCO Networking Technologies

Now it is time for us to get started on some of the different companies and techniques that we are able to use when it is time to handle some of the work that we want to do with our computer networking. And we are going to start this out with some of the information that we need to know when it comes to Cisco networking.

Cisco Systems is one of the leading networking companies that is known the best as a manufacturer and vendor of the equipment that we need for networking. The company is also able to provide us with some of the software that our networks need and offers other related services at the same time. Over the history of this company, Cisco has been able to focus on the IP or Internet Protocol options that we talked about earlier, and most of the networking technology that use this protocol as well.

This company doesn't just stop here though. It is also able to work with things like the routing and switching of products. Along with some of the technology that helps us with our home networking requirements, it helps with some optical networking when necessary, IP telephone, security, wireless technology and storage area networking for companies and small businesses as well.

If you are looking for some of the products that are needed to help with your networking, and you want to make sure that you handle some of this in a manner that is fast, efficient, and will be able to handle a lot of the work that you would like in your network, then this is the right company to look for. They specialize in pretty much all of the things that we are going to need when it is time to handle our computer networks, whether we are working with a small business or home network or if we need something that is larger.

Sandy Lerner and her husband Len Bosack were the ones who founded this company in 1984. At this time, Bosack managed some of the computers that were a big part of the computer science department for Stanford University, and Lerner did the same job but was at the Graduate School of Business at Stanford as well. It was during this time at Stanford that Bosack, along with Kirk Lougheed, developed the first product in Cisco, which was the AGS router.

With this in mind, we need to take a look at some of the networking technology that is available with the Cisco network, as well. This is going to help us to see some of the ways that this is all going to change over time as well, and how we are able to use some of these technologies in order to get the best results as well. Some of the things that we need to consider when it comes to our networking and ensuring that we get the most out of the whole process as well.

Why Machine Reasoning is Better than Machine Learning

Much of the business world right now is focusing on machine learning and how it is able to benefit them. And there are a lot of reasons to consider working with machine learning to help you create some amazing algorithms and to ensure that you get a good look into your customers, what they are hoping to get out of your business, what they customer is getting out of your competition, and more. But when it comes to the Cisco company, they wish to take a slightly different approach in all of this along he way as well.

For example, right now, Cisco is going to direct a lot of the money that it spends in the development department on some machine learning algorithms. This is because they would like stuff networks with all of the intelligence that they need in order to secure and manage them as billions of IoT and mobile devices are likely to come online and need to use this data, in just the next few years.

But this company also recognizes that something that is even more potent compared to machine learning is machine reasoning, which will promise us some algorithmic software that is capable of performing tasks and debugging problems like network engineers are able to do today.

We need to spend some time taking a look at how this machine reasoning is different from machine learning. To keep it simple, machine reasoning means that we are going to be able to program a computer with some techniques of logic, including induction and deduction. The machine reasoning system is going to come with an inference engine that is going to be able to draw some intelligence from the relationships that we see between rules and facts in a specific domain. This domain could be something like an area of network management.

The information that we are able to load up on a system of machine reasoning is often going to come from the practices that we notice with the experts in a particular field. Computer scientists can develop some great ways to turn this data into some kind of language, one that the computer is able to read through, and then the computer can take that information to draw some intelligent conclusions at the end.

Those who like to work with machine reasoning and use that for some of their networking technologies will argue that this is a more effective method to use compared to machine learning. This is because the machine learning will use some algorithms in order to perform statistical analysis on its own on facts that come up with some of the recommendations out there.

The computer is able to put forth some effort to debug a problem in the same manner that experts would do. When it works with machine reasoning, it is able to do a much better job at identifying some of the root causes of the problems, in a manner that is easier and more efficient than what we are going to see when we just choose to rely on machine learning.

Some of the more advanced technologies that are out there, similar to what we are going to see with machine reasoning, will become critical to engineers who are trying to troubleshoot networks serving numbers of devices higher than they are today. In fact, the IDC predicts that when we just look at how many IoT devices are connected to the internet, it is believed that this is going to grow up t 29 percent a year, reaching up to 42 billion in just a few more years, by 2025. Together, this means that the devices are going to hold onto about 79 zettabytes of data. And machine reasoning and some of the networking technologies that come with it are going to help drive this data, help these devices work faster and more efficiently, and so much more.

There are many companies interested in working with machine learning. And this is definitely something that your company could benefit from. But it does not mean that you need to automatically go through this process. It works great for some but not for all.

Learning whether machine learning is right for you will determine if you should work with it or not. Machine learning is expensive and takes a lot of time. And depending on your business, your goals, your customers, and more, it is possible that machine learning is not for you. You may have another option that is actually the best for you.

Computer networking, with the help of the technology found in the Cisco company, can help us to make this decision. When we understand how our business runs and how we can best find the right tools that will improve our business.

Optical Networking

Another option that we will see when it comes to taking care of the work that we want to do with computer networking and Cisco is the use of optical networking. In December of 2018, Cisco bought the silicon photonics company known as Luxtera. This was something that cost them $660 million and it helps to show that Cisco is trying to make a beeline to some of the work that it can do with optical networking. The transceivers that come from Luxtera and are used for data center routers and switches will be able to turn the electronic traffic into much faster light beams when we run them through the optical fiber compared to some of the other options.

Think about how much faster your networking is going to be overall if you were able to get the messages and the data across in no time, no matter how big or small those are. This is some of the potentials that we are going to see when it comes to optical networking, and it is a big area where Cisco is trying to improve its own technologies as well.

For many companies, data is going to be everything. And communication with speed is so important as well. If we are able to send large amounts of data throughout the world in seconds, rather than minutes or hours, it can make a big difference.

Think of it this way. If you work in the stock market and consider yourself a day trader, you need to move fast. If you see a change in the market one moment, you will want to make the purchase or the sale right away. When there is a delay in the system, then you will end up taking too long and losing a lot of potential money along the way.

With some of the new technology from Cisco, this issue can be reduced. Your transactions and communications can be almost instant. This helps you, not only in the stock market but also with some of the other places where you need to work with things moving quickly and time being of the essence.

The interest that Cisco shows in this optical technology is eventually going to allow them to use it in a lot of different manners, including in the application-specific integrated circuits and line cards. Without the faster speeds in data at some point, both of these technologies cause bottlenecking of the data. This is because the centers that are working with the data will start to struggle in order to keep up with the increasing amount of data that goes back and forth between all of those networks.

The widespread optical networking could be a great thing for many users. For example, it could be what will contribute to making multi-user augmented reality more of a mainstream idea that we see show up in more businesses and technologies in the future. Today, while it is the method that we see in work in some situation, it is going to be too slow for many people to see and interact with the same content of AR, or augmented reality at the same time, and this will likely cause some issues along the way.

However, if these optical networking capabilities are being used to the full potential, in the future it is likely that we will come up with a wireless network that has low-latency, high-performance, and high-quality along the way. This will help us to get all of this to many devices, even on the same network, at the same time. This allows for the use of AR for as many people as we would like.

There is a lot of potential out there when it comes to working with AR and what it is able to help us accomplish in the long time. For example, students are able to benefit from this augmented reality when they use it to visualize some of the scientific and mathematical concepts that they need while also providing corporate workers with some better diagrams to use while repairing the equipment in-house or for their customers.

CBRS and Cognitive Radio Emerging Networking Technologies

There is a lot of optimism out there when it comes to the trends that the communications industry is going to deliver when it comes to efficiency and speed with the radio spectrum management. We have to remember before we dive in that this is seen as a finite and shared resource, so effective management is one of the only ways that we are able to serve and satisfy the billions of devices that are found around the world that will be on and start to rely on this kind of network.

Some of the technologies that are the most likely to help make these future networks more robust will include options like cognitive radio and the Citizens Broadband Radio Service, but it is likely that this will expand out to more options as time goes on. In fact, it is believed that the industry, using some of the computer networking technologies that are being developed by Cisco, will be able to deliver cognitive radio advancements, will make it easier for us to significantly increase the number of devices that are going to be found on our wireless network. This cognitive radio is one of the form of wireless communication engineers that will be used to help plant the necessary intelligence into the needed transceiver.

The smart device that we are using with this one is able to search around and find some of the communication channels that are not in use at the moment. And when these are found, then the smart device is able to move the traffic over to those channels, keeping the traffic moving and ensuring that it is going to work in the manner that it should. The result of all of this is that there is going to be an optimal use of the available radio frequencies, and a minimal interference among the users of the network.

There are already many devices found on the radio channels around us. Each time that we get online, even on our phones, this will hook up to the wireless network and need to broadcast the signal. As the numbers are not likely to go down in the near future in terms of the number of users on the network, an issue with latent services and slow reception could turn into a big problem.

No one wants to spend a lot of time waiting for emails to go back and forth. No one wants to worry about whether they will have reception and be able to complete a call. We have come to rely on the networks that we use on a regular basis. And luckily, with some of the newer technologies available with computer networking, we can rest assured that our service will stay strong.

The idea here is to make sure that the radio frequencies from all of these devices are spread out and shared on multiple networks, not just on one. This system will help us to spread out the channels, and move them to frequencies that are not as highly used, resulting in better networking along the way.

As time goes on, and more devices are added to these networks, and more data is shared over the network as well, this is going to become more important. The faster that we are able to get things connected and moved over to the right channels, the better. No user wants to have a lot of drag time when they are using their device and trying to get onto the network, and when this is set up to work in the right manner, it is not going to cause as many problems, even when the number of users does go up.

As we can see, there are a lot of things to consider when it is time to work with the idea of networking and making sure that our devices and our computers are going to be able to hook up with one another and see the best results overall. And with the help of some of the Cisco technologies that are out there, we are able to make all of this work and line up in the manner that we would like as well.

Chapter 9: How to Handle CCNA and CCENT

Another topic that we need to spend some time on will be the CCNA and the CCENT. These are similar topics, but they work in different manners, and provide you with some of the certification that you need when you would like to work on some of these networks. Let's dive in and see how these work and how they are going to be different from one another.

First, we need to look at what each of these means. The CCNA is going to stand for the Cisco Certified Network Associate, and then the CCENT is going to stand for the Cisco Certified Entry Network Technician. Both of these are certifications that belong to the ecosystem of Cisco, which is a mass provider of some of the components of infrastructure and networking that we need on a regular basis. This company is also known for being a pioneer in some of the innovations that we find with networking technologies, routing, and switching over the past few decades.

However, even though both of these come from the same kind of ecosystem, we need to be careful about how we are using them, and we need to see how they are different from one another. For example, the CCNA is a certification that will expect the students to prepare and administer some of the medium-sized routed and switched networks that we may choose to work with. But we can take this to the next level with the CCENT certification because it is the most advanced out of the two.

IT certifications are often some of the best and the most common, methods that we can use in order to demonstrate the skills that we have to employers, and can ensure that you are hired for the job that you would like. And when you can combine it together with the CCENT and CCNA, you will find that your skills are going to be some of the most in demand that are out there.

There are also a lot of certifications out there in various domains of this sort, such as when you want to spend time testing, you would want to work with the option of ISTQB, because it is one of the most recognized. And if you would like to show some of your skills and experience with the Cloud, then you would want to work with the AWS Solution because this is going to show that you are good with working in Spark, Hadoop, and Big Data.

These certifications are helpful because they teach different skills at a wide variety of levels, and they are directed towards different job types and IT positions. Students will need to spend a good deal of time learning before they even show up for the exams, and they are going to need to be renewed once every three years.

The CCNA is going to expect that students are able to prepare and then administer on the medium-sized routed and switched networks that are there. They are also going to cater to the WANs, which are used in this certification to help establish the connection between devices and computers in a wide variety of geographical locations.

Students who have been able to complete this kind of certification are also aware of any and all preventative measures to help with these connections, and they will be able to do some of the necessary troubleshooting on basic security. They know how to keep the network safe, and what needs to happen to make sure that a hacker or anyone else is not able to cause a threat on the network and steal all of the data that is on that network.

Now, working really in-depth about the security of a network is often above the CCNA, though knowing the basics is important. And that is why it is important to work with the CCENT certification in order to get some of the more advanced options that you need. One thing to note though is that it is not required that you take the CCNA before you get the CCENT. Some people get one before the other, but you can definitely skip right to the CCENT if that one tends to meet your needs a little bit better.

Both of the exams valuable to many of the employers that you would like to work with, and can show that you have some of the practical and real-world skills needed for doing networking with Cisco Both of these also show that you can handle a variety of issues, no matter what the business settings are in the first place. These are really helpful if you are trying to get into this kind of job field and you want to make sure that your application stands out from all the others.

There are a few things that we should consider when it is time to work with both of these. First is important to note how different the two of them are from one another. They will share a lot of similarities in many cases, and they will both show that you have some good experience when it comes to working with computer networking, but it is important to know some of the differences so you know which of these to work with.

The biggest difference that you see between the CCENT and the CCNA is related to the depth of the scope of what you will learn. The CCENT is based on more of an entry-level design that is there to serve as a good starting point for any professional who would like to enter into a new career that focuses on the field of networking.

The topics that are found on the CCENT are basic and while you do need to study and be prepared, they are not as in-depth as some of the other certifications that you would need to do. Some of the topics that you will find with this certification course include installation, operating, and supporting some of the small branch networks. This is a great way to get the basics of the world of networking down and can help you when your career is going to focus on the small and the home networks that you can find.

On the other hand, we can also work with the CCNA. This one is a bit different because it is considered more of the associate level of the examination and the curriculum and the courses that you may take to get this one is a bit more complex compared to the other options. If you would really like to get more into some of the work that you are doing with this process, and you want to really stand out with what you are able to do, then this is the certification that we want to work with.

You will find that it serves as a good value to add to some of the foundational knowledge that we are able to cover in CCENT and so if you have been able to do the CCENT option, then you are set up and ready to go when it is time to get started on some of the curriculum that you need to do with the CCNA.

Keep in mind with this one though that the skillset, depth, and scope that is required in order to get through the CCNA examination are going to be a lot harder, deeper, and higher than we are going to see with the CCENT. Topics, including the advanced dynamic protocols that are going to be used with routing, VPN, and even Rapid Spanning Tree or RST, are just some of the topics that we are going to cover when it comes to working with this process.

In addition, you will find that the CCENT is not only aimed at giving us a good understanding of what these networks are and how we work with them but will make sure that you know how networking relates back to other applications and operating systems as well. For example, a database administrator, which is going to be a programmer of computer software, and a system administrator may decide that it is a good idea to learn more about passing this CCENT exam so that they gain a full understanding of these networks and the impact that they are going to have on getting the system to run the way that we would like.

Now, we have to remember that there have been some battles that happen between the system admin team, the programming team, and the networking team in some cases, especially when we are talking about where a bit IT problem is going to be found. Having a better understanding of the networking is going to help us to minimize these instances because the whole team, no matter which department they are in, will be able to understand how this works and what we are able to do with it. The CCENT exam is often going to help with this.

While you are not required to go through and do one exam before the other, the CCENT exam is often going to be seen as one of the first steps before you work on taking your CCNA certification. This will ensure that you are prepared, that you have everything lined up and ready to go, and that you are able to really start to understand how these networks are going to behave before you dive in any more to the process.

If you are trying to learn more about these networks and what we are able to do to make them work for our needs, then taking one of these certifications is going to be the best course of action. It will ensure that you will really understand what is going on with the network and that you will be able to handle some of the problems, the connections, and more that are going to come up with them. And that is always going to be something that looks appealing to any employer you would like to work with now or in the future. If you are ready to work with networking and to take the knowledge that you get from this guidebook and put it to good use, then this certification is going to be the next step to help you do that.

Chapter 10: Introducing the OSI Model

The OSI model or the Open System Interconnection mode, is able to help us define a networking framework so that we are able to implement the necessary protocols, and it is going to do this in seven layers We need to spend some time looking at this model, learning what it is all about and what it is going to help us do, and see how the seven layers, which we will talk about in more detail later on, work with one another to get things done.

What is the OSI Model?

The OSI model is able to define the networking framework that we would like to use to implement these protocols with the help of seven layers. There isn't that much to the OSI model, which will make it a bit easier for us to work with in the long run. You will find that it is not even tangible anytime that you decide to work with it. The OSI model, for example, is not going to actually come in and perform any of the functions that are needed to complete our networking. It is more of a conceptual framework, which it will help us get a better understanding of some of the complex interactions that are showing up around us.

There are a lot of great models that we can work with, but you will find that the OSI model is one of the best options to spend time on. But you can always spend some time looking at some of the other models along the way to ensure that they are going to behave in the manner that we would like.

This brings us to the idea of who developed this OSI model. The ISO or the International Standards Organization developed this model for us to use and to give computer programmers a simple and basic model to get things done with. It also helped us to divide the communication that we see on the network into seven layers.

In this model, we see that the layers 1 to 4 are considered some of the lower layers, and they are the ones where we spend our time moving data around to meet our needs. Then we work with the layers 5 to 7, which are known as the upper layers. These contain some of the application-level data that we need to work with. Adding all of these together is important in this process because, in the end, this model is set up to move the data around that we need, and can easily show us the basics of any computer networking project that we want.

Another thing that we need to realize before we get into the seven layers of the OSI model is that our networks operate on one basic principle, which is the "pass it on" principle. This means that each of the seven layers that we focus on will take care of a job that is specific, and then it will pass the data over to the next layer and be done.

The Seven Layers of the OSI Model?

We have mentioned them a few times in this chapter already, so now it is time for us to take a look at some of the layers that show up with the OSI model. The OSI model will focus on having all of its control passed from one layer over to the next. This starts with the Layer 7, which is going to be the application layer in one of the stations, and then will proceed to the bottom layer, over the channel to the next station until it makes its way up the hierarchy. This model is going to help us take the task of inter-networking and will divide this up into what will be known as the vertical stack that will have the following 7 layers that we would like to work with.

We are going to go into these seven layers in more detail but first we will list them out! The seven layers that we see will be the application, presentation, session, transport, network, data link, and physical. Let's dive into each of these in a bit more detail to see how they work for some of our needs as well.

First, we start with Layer 7, which is the application layer. This layer is there to support the end-user processes and all of the application. Communication partners are identifiers, and we will look at the quality of service and find out whether it is good or not. The user privacy and authentication are something else that this layer will need to consider, as are the constraints that may be present on the syntax of the data as well.

Everything that happens at this layer is specific to the application that you are working with. This layer is to provide us some services to help with e-mail, file transfers, and some of the other network software services that we need. FTP and Telnet are applications where the whole of them found on the application level. Tiered application architectures are part of this layer as well.

Then we move on to Layer 6, which is the presentation layer. This is the layer that will provide us with some independence from the differences that are found in data representation. Often this is known as the process of encryption, which is where we translate from the application to the network format and vice versa if we would like.

This layer is important because its job is there to transform the data into a form that our application layer from before is able to accept. This layer is also a good one to use when we would like to format and then encrypt our data to go across the network, which provides us with some freedom from compatibility problems. In some cases, this is known more as the syntax level as well.

Next on the list is Layer 5, which is the session layer. This is an important layer because it can establish, manage, and then terminate the connections that happen between the applications. The session layer is also going to help us by setting up, coordinating, and then terminating any of the dialogues, exchanges, and conversations that happen between applications at each end. It also deals with the session and connection coordination.

Now we are on to Layer 4, which is the transport layer. This the layer in the OSI model that will provide some transparent transfer of data between the end systems or the hosts. It is also responsible for the end-to-end error recovery as well as the flow control. And it will ensure that we are able to complete the transfer of the data as we are using it.

We can also work with Layer 3 in this model, which is the network. This layer is the one that will provide us with the technologies of routing and switching. It can also create some logical paths, which we are going to know as the virtual circuits, while also being good for transmitting data from one node over to the next.

This layer is also going to spend some time with the routing and forwarding of the data. But this is not where we stop with all of it. This layer is also used to help with error handling, internetworking, addressing, congestion control, and even with packet sequencing. There is a lot that will show up in this layer, so it is important to learn how this one will work for our needs as well.

The next layer that we are able to work with is Layer 2, which is for Data Link. When we are looking at this second layer in the OSI model, we will notice that the data packets are encoded and then decoded into bits. It is going to furnish some of the knowledge that we need about the transmission protocol, and will be able to manage and handle some of the errors that happen in the physical layer, some of the issues with flow control, and is good at frame synchronization.

This layer of data link is also able to be divided up into two other sub layers. This is going to include the LLC, or the Logical Link Control, or the MAC, or the Media Access Control layer. The MAC layer is important because it is in charge of controlling how a computer on our network will be able to gain the access that it needs to the data, while providing the necessary permission in order to transmit that data. Then the LLC layer is going to control some of the other parts like error checking, flow control, and frame synchronization to name a few.

And finally, we reach the Layer 1 of the OSI model, which is the physical layer. This one is similar to a stream. This is going to include a stream of radio signals, light, or even electrical impulses, through the network, and this is going to happen at the mechanical and the electrical level. It is also going to provide us with some of the hardware that we need in order to send and receive data on the carrier that we are using, including helping us to define the cables, physical aspects, and the cards. Some of the examples of these physical layer components are going to include ATM, RS232, and Fast Ethernet.

All of these parts need to come together in order to help us to work with the OSI model and to ensure that the data is going to work in the manner that we want. When we put all seven of these layers in place, it is easier for the networking to happen, and the various parts of the network are going to be able to come together and actually communicate and do what we would like in the process as well.

Chapter 11: A Word About Network Security

With the increasing amount of reliance that most of us have on technology, it has never been more important for us to go through and secure all of the aspects of our online information and the data that we put on our network. And as we see the internet grow more and that the computer networks around us are getting bigger as well, the integrity of our data is really one of the biggest considerations that an organization has to worry about. Because of these issues, and more, it is important that we spend at least a bit of our time worrying about network security and how it is going to affect us.

Why is Network Security Important?

To start with, we know that network security is going to be really important to our overall goals. This is going to be even more important if you plan to bring your network online and use the internet at all, which is pretty much a guarantee in this day and age. It doesn't matter how big or small your network is, if you are online and sharing information with others, then it is important to worry about the security of your network as much as possible. Having a good network security system in place, no matter what, is the key to ensuring that you are able to reduce the risk that you have of falling prey to a hacker, data theft, or some other form of sabotage along the way.

There are a lot of benefits to working with your network security, as well. For example, it can provide you with some good protection against harmful spyware that a hacker may try to get onto your computer or even the network as a whole. It is also going to ensure that any of the data that you are sharing with others on the network will be kept as secure as possible.

The infrastructure that we find with network security is going to provide us with more than one level of protection, in the hopes of preventing many attacks like a man in the middle attack. This is done when the infrastructure is able to break down the information into numerous parts, taking those parts and encrypting them, and then transmitting them through some of the independent paths that are necessary, thus preventing cases like eavesdropping and other things that a hacker may try to do against you.

Getting connected to the internet can bring about a lot of benefits, but it also means that you are going to get a ton of traffic coming in and going out as well. This huge amount of traffic, when it is not monitored, is going to cause some stability problems and could lead to some vulnerability issues in the system. When we work with network security, it is going to promote more of the reliability in our network because it prevents lagging and downtimes, simply because it continuously monitors any transaction that seems suspicious and might end up sabotaging the system if that transaction came through.

What Can Go Wrong On My Network?

It is never a good thing to have your network hacked no matter what the situation is or why you use the network. This is going to lead to a really big mess that can take years to fix. And if it ends up in a big data leak, it can ruin your reputation, cost you a lot of money, and make some big headaches for all who were a part of that network, even those who shopped with you in the past.

Having any part of your network hacked is detrimental, and it can end up putting you out of business. Vandalism is one of the things that can happen when the hacker does manage to get on your system. This vandalism of the network is where the hacker is going to plant some information that is misleading on your system. But this is just one of the many tactics that a hacker is able to use against you. When the hacker is able to plant this wrong information though, the integrity of your company can be called into question, and it is possible that many of your customers can be misled and harmed in the process as well.

Another impact that we have to be careful about when it comes to the security of our network is when a hacker works to damage intellectual property. Hacking is going to give access to the hacker that is not authorized to the individual's or the company's information. And when the hacker gains this information, they are able to do whatever they would like with it.

For example, we can take a look at the security breach that happened with Citibank a few years back. This ended up being so large that it was estimated that about 1 percent of the U.S. customers for this company were affected in some manner, and it took forever to clean up that mess. If a hacker is able to get in and steal the ideas, plans, or blue prints of what your company would like to do in the future, then it is possible that your company is going to be left out and will not be able to implement some of the new products and designs that it would like. This is never going to end up with something good whether it stagnates or destroys the business in the process.

The company, as well, will often experience a loss in revenue when a hacker is able to get onto them. Most of the attacks that a hacker is going to launch on the network are going to lead to crashing. And in the extended downtime, you will no longer be able to make sales and will lose revenue. The longer that the hacker is able to keep your network down, the more you are going to lose in revenue. And as you start to look unreliable and lose your credibility, this can become more of a long-term problem that you need to worry about.

This is not the only way that a company could lose some revenue, though. It is possible that the hacker could get on and steal financial information when they get onto the network, which can lose the company a ton. And if the hacker is able to get onto the personal and financial information of the customers, the company may have to pay to fix the issue, pay the customers, and pay fees for not maintaining their network security. This is one of the main reasons that so many companies are going to fail when they have one of these data breaches.

Even a personal computer can be taken advantage of. In fact, there are many personal computers that fall prey to this on a regular basis. Companies are good at protecting the valuable information they hold onto. They know that only one hack can end up costing them their whole business and could be the financial information of hundreds of thousands of people at risk.

These companies know the risk, so they usually put up all the protections they can to keep hackers out. These companies have firewalls, anti-virus, antimalware, and many rules that employees need to follow to prevent social engineering attacks. They may also hire teams of security personnel to watch their data and their databases, and ensure nothing gets in.

Yes, there are some attacks, but often it is not because of negligence on the part of the company. They may have done everything in their power at the time to keep the hacker out.

This usually results in something out of the control of the company at that time, even if it is detrimental.

However, if you have your own personal computer, you may not put as much time and attention into the security of the network as you should. Too many users of personal computers assume that they don't have anything of value for a hacker, and so they forgo some of the security measures that they should.

The truth here though, is that if you ever get online, do a search or send emails or anything else that needs a network to accomplish. Then you need to take the security of your system seriously.

Hackers know that personal computers are easier to get through. They know that individuals do not put on the same protections as companies do to their networks. And while it is hard to get onto a database of thousands of users from a company, hackers know that it is much easier to hack into thousands of computers with little to no security.

While your computer and information may not seem like much, when it is added to a list with thousands of others, it does become very valuable. And before you know it, all of your information, your identity, and your security are gone.

Even if you are the only one who uses your network, and you are certain that you have nothing of value on the system, network security is going to be important here. You need to take pride in the work that you do and keep yourself safe, with the right protocols, careful consideration of any network you get onto, and even a good security system that includes a firewall and anti-virus tools.

It is important for us to spend some time paying attention to the data that we have, and the security of our network. If you are working with customers' it is your duty to make sure that their information stays safe. And if you don't do this, it is going to cause a big mess that you then have to clean up on your own. Taking the time to keep the system up to date and worrying about the security of your network is going to be so important for helping your business to grow, to keeping the trust that you need with customers, and more.

Conclusion

Thank you for making it through to the end of *Computer Programming – First Steps* let's hope it was informative and able to provide you with all of the tools you need to achieve your goals whatever they may be.

In this day and age, technology is everywhere and part of every thing. Having an understanding of what goes on behind the screens will undoubtedly help you, no matter your path. The next step is to get started with some of the suggestions that we talk about in this guidebook. The information given here will help you when it is time to work with this process. It is important to truly grasp the process if it will serve us in our future endeavors.

While a lot of people are going to spend their time just using their devices and their computers without really thinking about the mechanics that go into it, having this kind of understanding is really going to help our progress and makes us appreciate some of the things that our computers do a little bit better. This understanding will help you succeed in a number of different parts of your life.

Success is the main goal of working with this guidebook. Learn about being able to set up an IP address, understanding some of the technologies out there, the importance of working with networking, and some of the options that you can choose based on the size of your network. This is all going to come together to help us see more about the different parts of our network, and how they are all about to work together to make the communication and sharing of data better than before.

Many times learning more about your computer can be beneficial. This is true whether you are looking to just improve your own skills in computer programming or if you are interested in learning more about some of the mechanics that are going to show up behind the computers and the devices that you are going to use on a regular basis. When you are ready to learn a bit more about computer networking and what you can do with all of this information, make sure to read through this guidebook to learn more about how to get started.

Finally, if you found this book useful in any way, a review on Amazon is always appreciated!

CPSIA information can be obtained
at www.ICGtesting.com
Printed in the USA
LVHW050904250121
677402LV00011B/429